I Don't Speak Geek

A Simple Guide to Help Businesses Navigate Today's Complex Technology Choices

Teryl L. Burt

ISBN-13: 978-1456378608

ISBN-10: 1456378600

Printed in the USA

SPECIAL THANKS

Thank you to my sibs, Gloria and Herb (we are often known as the Burt Brothers) for a 32-year unshakeable partnership.

Thanks to my mentors, (in the order in which I met them), Chuck Grant, Bob Farkas, Christine Anderson, Oli Thorardson and Kent Erickson.

RR, you know who you are—thanks!

And thanks especially to an amazing array of loyal Clients who have taken the wild ride with TSG through three decades of rebellious growth, to have finally reached technology infancy.

How to Read This Book

Not too slow, not too fast. Kind of half-fast.

Louis Armstrong, jazz trumpeter

This book is way too long to read (unless you really want to be geekful).

I suggest scanning the Table of Contents to find the issue that's been bugging you, the subject of your fourteen year old's endless jabbering, something about which you are uncomfortable asking out loud or just an area that intrigues you.

Little in life is for everyone but if you are forced to manage technology in your business, school or non-profit organization, there might be something helpful in these pages.

You may also find something interesting in the "tip" blurbs. Finally, if you are a practical sort, just read the Secrets and the Case Studies.

Table of Contents

TABLE OF CONTENTS, CONTINUED . . .

INTRODUCTION

jar·gon [jahr-guhn, -gon]
–noun

1. the language, especially the vocabulary, <u>peculiar to a particular</u> trade, <u>profession</u> or group.

2. <u>unintelligible</u> or <u>meaningless</u> talk or writing; gibberish.

3. any <u>talk</u> or writing that <u>one does not understand</u>.

4. language that is characterized by <u>uncommon</u> or <u>pretentious</u> <u>vocabulary</u> and <u>convoluted</u> <u>syntax</u> and is often vague in meaning.

My soon to be 11-year old friend still giggles when I ask him if he has triskaidekaphobia. For years we had great fun as I tried to stump him with weird, complicated words with funny meanings.

To a business owner, weird complicated words are not funny. They are useless, time wasting and often insulting. Those who try to dazzle their potential employer with the technology jargon, dubbed *Geek Speak*, will find themselves without an audience in the quiet of their own garage, expounding on 4200 rpm spindle speeds and rotational latency to create robust architectures.

Considering the plethora of technology options to choose from today, it is no surprise that advice is all over the map. What this book provides is simple explanations of options that will lead the folks at the helm of businesses, non-profit organizations and educational institutions to optimal technology solutions based on their unique requirements.

CHAPTER 1

RUN A BUSINESS? WHY _YOU_ _NEED_ TO READ THIS BOOK

"Virtualization emergence paradigm shift broadband SaaS network sync connectivity hypervisor bandwidth HaaS task-based dedupe software bloat spybot Cloud.   Thank you."

HUH?

If you own or operate a small or medium sized business (SMB) today, you know that you need technology. You need it more than ever, not only for competitive advantage but just to keep your boat afloat. The younger people on your staff don't even know what it was like not to have email—it has become such a business critical service that tolerance for outages is close to zero.

Navigating the technology maze called "infrastructure" can be a painful process, whether you have a general idea about what you need, or haven't a clue how information gets through the air from A to B.

Many SMBs survive by piece-mealing their automation. The owner hears about some software that allows faster invoice processing. The software is expensive but hey, invoices would get paid much faster too. The software is purchased, another piece of expensive hardware is purchased to run it on and voila: It's a system.

But darn, the printer won't work with this software and the phone line is too slow to download the frequent software patches and upgrades. The accounting computers are spewing garbage so anti-virus software is brought current and a consultant comes in to install it and clean up the workstations.

The consultant informs the owner that the hard drive on the file server is much too small and wants to know who spec'd out such an ill-fitted system anyway. Bigger hard drives are purchased on sale at Best Buy but they don't fit in the server, which is proprietary and uses only parts manufactured by that same (4 letter) manufacturer.

The list of **sunk costs** grows and grows as incompatibilities are uncovered, security threats require more purchases (like a firewall and spam filtering software), licenses must be purchased, constant system maintenance (like patches) must stay current . . . until finally the original equipment has reached the end of useful life before the business owner produces a single fast invoice.

More than likely, you are not interested in technology or you would have entered that high falutin' profession instead of starting a business. The truth is, you went into business because you are good at something or you enjoy something or your parents left you the business, which you know inside and out because you were forced to work there every summer. You want to focus on that something and make more money or serve more people or animals or create a legacy for *your* children.

You do *NOT* want to futz around with stuff you may or may not understand. You do *NOT* want the expense of an outside consultant or an employee who rattles off a bunch of jargon-rich techno-babble, insults your intelligence and confuses everything.

Unless your favorite daughter-in-law is an Information Technology (IT) Director with spare time, you better:

A. Learn to speak Geek

or

B. Find someone to trust who will help you make informed decisions, look out for your wallet and keep you out of the fray so you can do what it is you set out to do:

Focus on running and growing your business.

This book is about demystifying the options for business owners to use technology effectively and minimize costs.

The Three Most Common Complaints About IT Professionals:

1) **Slow Response Time:** Without quick response, a down or problematic network can bring an entire office, store or organization to a revenue ravaging halt. If this is your complaint about your IT support, move on (See Chapter 9). It is likely an omen of what is to follow.

2) **Poor Communications:** Technical sorts are legendary for poor communications skills. This is no longer acceptable—you should not have to translate Geek Speak to run your business.

3) **Lack of Follow Through:** Guarantees and commitments about seeing projects through to your satisfaction are called "SLA's," for Service Level Agreement. These should be set or negotiated when entering into any new agreement.

Chapter 2

Top 10 Reasons You Know Something's Gotta Change

If you find even a few of the statements below to be true, it is time to start considering a change to how you manage your technology.

1. Your Information Technology (IT) spending is willy nilly and out of control and you would prefer to work with a budget.

2. Your hardware needs "refreshing" but your inner voice (channeling your dad) says, "Protect the Cash."

3. You pay the big IT Salary but your IT person(s) never has time to do anything but Help Desk.

4. You have a hard time focusing on the "top line," worrying about bottom line costs.

5. You worry about risk: what if my network goes down and we are out of business for hours, maybe days? What if our backup isn't working and we lose data?

6. You know your software licensing isn't current but don't know how to figure it out and can't afford to upgrade it all at once anyway.

7. You have heard about the trend toward "outsourced support" but your IT staff doesn't know much about it, is threatened by it and so calls it a passing fad.

8. Your IT person(s) has 26 weeks of accrued vacation because you are afraid to be without coverage, for even a day.

9. Your insurance company asks you where your documentation is kept for Disaster Recovery purposes and "in so-and-so's head" will not get you that discount.

10. You want to run your business, not manage drudgery. Period.

Running a business is a challenge in the best of times. In today's difficult economic climate, it is even more critical to the mission of the business or organization to use technology dollars wisely.

Business as usual went out the door with the financial crisis but companies of all sizes still need efficient business processes and solutions. They need to figure out how to optimize what they have, what they should purchase and the best use of their hard-earned dollars.

To **Think About:**   Managing your business *is* your business.

If managing technology is distracting you from your core purpose, read on.

CHAPTER 3

HELP! I WANT TIME TO RUN MY BUSINESS!

If you are short on skills in-house, need to augment your internal capabilities, or you want to focus more on running and growing your business, there are multiple options available for getting outside support.

The practice called "IT Outsourcing" refers to hiring an outside service provider to perform a traditionally in-house IT function(s). IT Outsourcing (not to be confused with using off-shore services) is an increasingly popular strategy for reducing the cost of IT while, in most cases, improving the service you receive. The service provider organizes and manages the work and takes responsibility for the outcome, to the great relief of many business owners.

Some IT services lend themselves perfectly to outsourcing (like email management, disaster recovery and security management) because they are complicated and expensive to manage. Others (like desktop support and server support) are appropriate because the work is routine so it can mostly be automated and handled remotely, making it comparatively inexpensive.

As is the case with most successful business relationships, a referral to an outside service provider from a friend or colleague you trust and respect is a good place to start. You are looking for a <u>neutral advisor</u> who will work with you to recommend the most cost-effective and efficient support option(s), <u>not just what that particular vendor sells</u>.

Factors to consider in choosing how you work with your outside IT vendor include:

- the need for mission-critical response time (retail stores need it, medical facilities need it, banks need it, lawyers sometimes need it; most businesses can survive short wait times to reduce support costs);

- the amount of work likely to be required in the coming year for emergencies, routine maintenance, growth and upgrades;

- the level of your company's in-house expertise (think hard about this--some upgrades, like a network file server operating system upgrade, are only required every 3-5 years. Whoever completes this type of upgrade should have very fresh current skills).

The next section describes both classic and emerging options for ways to work with an outside IT support vendor.

Real Logical Advice: Buy "outcomes," or results, not pieces. There are many pieces of the puzzle that can best be designed and assembled by an expert to be sure you end up with a unified solution that truly meets your business needs.

It may seem more costly on the front-end. It probably is not because of the multitude of conflicts you will avoid.

Explore Your Outside Support Options

Classic:

➤ Incident Support (Time & Materials or T&M)
➤ Block Time
➤ Project Outsourcing
➤ Fixed Bid

Emerging:

➤ Managed Services
➤ The Cloud

Classic Options

➤ Incident Support (Time &Materials)

Incident Support, or T&M, is typically chosen by a business with a single issue to resolve or a requirement for specialized expertise (such as data base programming or a new software application installation). Sometimes there is an unplanned emergency, such as a natural disaster or a utility outage and you need extra hands on short notice.

While some amount of Incident Support will always be necessary, businesses are moving away from this type of billing as part of a standard relationship. It is generally <u>the most expensive way to do business</u>.

To provide quality incident support the vendor has to keep hourly rates high because they have to keep skilled technicians and network engineers on-call to respond. The utilization rate for these technical people (meaning the actual amount of time the service provider can bill a customer versus how much they have to pay for payroll in a standard business day) can be very low, once you factor in traffic, wait time, and waiting for users to log off the server on short notice.

The vendor also has to pay auto expenses, bridge tolls and parking fees. In San Francisco, for example, parking averages over $30/day and bridge tolls are between $4 and $6.

The cost to the business is also high and not just monetarily—employee stress matters too. Everyone is forced to wait until a technician arrives, then all of the users must be coerced off the network (there are always one or two who need "just another sec" that turns into many more "secs"), there is resentment about missing deadlines (people generally want to do a good job) and everyone needs to figure out how to occupy their time while the computer technician is working.

Have you ever calculated what down time really costs your company? Consider the example on the next page. Then plug in your own numbers for some potentially startling results.

These are just the hard costs. Depending upon the nature of your business, you might also calculate the "Lost Sales Opportunity Cost," the "Lost Customer Cost" and the "Damaged Reputation Cost."

You can frequently reduce hourly rates by working under a Managed Services program (described in a later section).

TERYL L. BURT

Cost of Unplanned Down Time
6 Person Business

Average salary per employee	$50,000
Benefits at 20%	10,000
Total Ave Annual Salary/Employee:	**$60,000**
Work Hours in One Year	2,080
Total Average Hourly Salary/Employee:	$ 28.85
Number of Employees	6
Total Payroll Cost Per hour:	$173
Payroll Cost, wait hours tech to arrive (2)	$346
Payroll Cost, tech is working on issue (3)	$519
	$865
Hourly rate for technician @ $150/HR	$450
Travel time charge for technician	$ 65
Total Cost of technician for 3 Hours	**$515**

Total Cost of Unplanned, 3 Hours of Server Down Time: **$ 1,380**

The cost of unplanned down time to a business is enormous. Just a few of these can cripple a company. It is almost guaranteed to be "when" not "if" when working with technical support on a T&M basis.

➤ Block Time

Working under a Block Time agreement can be a cost-effective option for businesses that require some on-going support, but do not want a full-time in-house staff member or are unable to schedule their work on a regular basis (like, three mornings a week or every Tuesday).

Block Time is, as its name implies, a block of pre-purchased hours to be used wherever and whenever the need arises— help desk, emergency response, routine maintenance, upgrades, budget planning and review, expansion—pretty much anything technology related.

Block time hours are prepaid. This allows the IT vendor to reduce the rate (lower than Incident Support described above) and provide priority response time to the business. Often, the most compelling selling point for Block Time is priority response by someone who is already familiar with your environment—it can be a huge time and cost saver.

> **S**ecret: Sometimes a prescribed
> program does not fit with how you see
> your needs.
>
> IT Vendors usually have pre-set packages
> for specific blocks of time but if you do not
> see a plan that you think works, *ask*.
>
> An outside support provider should be
> able to figure out how to meet your needs,
> not ask you to change your business to fit
> their product (in this case, services).
>
> A great way to use Block Time is for
> vacation support for the harried IT person
> whose time off requests the business
> owner is always afraid to approve.

➢ Project Outsourcing (POS)

A Project has a start and an end, generally
defined by a Scope of Services document or
Letter of Engagement. IT service providers
can provide temporary, full time staff to
clients with specific needs, on a project
basis. Rates and terms are determined on
an individual basis, depending upon the
requirements and duration of the job.

Real World Example: My company works with a small medical office that decided to move to electronic record keeping. We were hired to "scope" or define the project and help the owner select the appropriate software.

The Client had rooms and rooms full of old records in cardboard boxes which they are required by law to keep for "x" number of years. Our job was to figure out how many records there were, how much electronic storage space would be required (building in room for growth), research available vertical (industry or business specific) software applications, spec out the equipment (i.e. high speed scanner that would integrate to their software selection) and finally, assist the doctor in demos with prospective vendors to make an informed selection of the software.

Our job ended here. In this case, the practice chose to do the scanning of the old records themselves with inexpensive high school students.

Working on a project basis can be beneficial to a business if it is well planned and documented. It is critical that the vendor provide a written plan, associated costs (usually in a range) and define what constitutes an Out-of-Scope incident which would incur extra charges. The vendor should be willing to work with a formal Change Order process that includes revised time and cost estimates, should something unexpected turn up, and the business owner should approve Change Orders in writing. This helps to reduce or ideally, eliminate billing disputes.

Really Important: The best predictor of a successful outcome is a good plan. If budgets are strained, planning is NOT the place to skimp. Better to have an experienced professional set the stage and work in phases or do some of the work yourself if you have to, than plan it yourself and risk facing disruptive gotchas later.

➢ Fixed Bid (FB)

On large projects, an IT service Provider may agree to work from a Fixed Bid. Pricing is determined by the scope of the

work. This is not the same as a "Project."
Both have a beginning and an end defined
by a Scope of Services document or Letter
of Engagement. The primary difference is
that the price of a Fixed Bid remains the
same, regardless of the amount of time and
resources it takes to complete the job.

Fixed Bid can be a risky way to go for the
service provider and is therefore often bid
at a higher price to the customer. Upon
completion of the job, there is usually a
winner and a loser. Either the vendor
finishes in less time than projected, in
which case the customer paid a higher price
than was necessary, or it takes longer to
complete the job than the vendor estimated,
and the vendor's costs are higher than
projected. Generally, it does not mean that
the customer got a better deal or more work
was completed than was anticipated. It just
means that for whatever reason, the
amount of time required to complete the
job was estimated incorrectly.

Practical Application For Fixed Bid **Pricing:** Network cabling is generally completed on a Fixed Bid by a licensed professional cabling contractor. This is an acceptable application for Fixed Bid because the tasks are routine and the Cabler knows precisely how many feet of cable, how many network cable ends, faceplates and other materials are required, based on a walk-through when the job is estimated.

Yet, even with an easily defined job like network cabling, issues can arise that seriously change the scope of the project. My company was once half way through a very large cabling job, spanning several different buildings, when we ran into a brick wall—literally. The hidden wall was two (2) feet thick. We brought in special equipment at that point, on our dime. The Client was as surprised as we were but we honored our FB price.

Emerging Options:

➤ Managed Services (MS)

The trend in IT is toward a more efficient and less costly method of delivering automated support services *remotely*, broadly called Managed Services (MS). These services are designed to provide enterprise class technical support to smaller businesses and organizations at an affordable cost.

The premise behind the Managed Service business model is that costs are lowered all the way around because it is a *pro*-active service, instead of a *re*-active service. Every aspect and component of your network can be monitored 24x7, so that existing or developing issues can be identified and fixed remotely, often before you are even aware of the problem.

Routine maintenance, such as software patching and virus definition updating, is performed on time, every time. Software patches are pre-tested and spam filtering occurs before anything reaches your internal network. Most of the support activities that used to require a potentially expensive on-site visit by a network engineer or technician can now be automated and handled remotely.

Managed Services programs are generally flat-fee so you are able to budget for the amount you spend. The costs can be materially less because the service provider's costs are less. The vendor is relying on the fact that well maintained systems are going to fail far less often.

MS programs range from simple network workstation monitoring and maintenance to complete, proactive management of your technology environment, including all other technology vendors (i.e., your Internet Service Provider (ISP)).

Businesses of all sizes often find that Managed Services programs are significantly less expensive and more efficient than managing IT internally.

The Value of Regular Patching:
Software manufacturers (Microsoft and most others) release "patches" to fix security vulnerabilities in your computers, fix bugs and improve features.

Feature enhancements may lack value for your business, but enhancing the security of your network (continuously) is *critical*.

Chapter 4 describes damage that can occur if you do not keep patching up to date.

The Cloud:

Way *More* for Way *Less*

Today many businesses and organizations work with a homegrown network. They create a detailed plan (hopefully!) to determine the business needs, purchase the hardware and software, build it themselves or hire a professional IT Services firm to install the pieces and add a high-speed internet line. Someone is hired or appointed from within (qualified or not) to act as the IT Support staff.

Now there is an emerging option to manufacturing your own IT, that provides centralized, (nearly) 100% uptime computing that is cheaper, faster and more agile than what most businesses can afford to do in house. It is called **Cloud Computing**.

There are public Clouds (like Amazon and Google) and private Clouds, generated and maintained by highly innovative, early adoptive services firms. Many businesses will find that a hybrid approach to Cloud Computing will best fit their needs and their budgets.

The basics of Cloud Computing are presented in more detail in a later chapter called **"1-800-The-Cloud"**.

Worth Considering: Even very small businesses *could* benefit from moving to The Cloud. For example, offloading email management to dedicated email hosting pros is cost-effective for almost everyone.

If your business has more than a few users, you should consider it.

An estimate is free, so why not?

CHAPTER 4

BUSINESS THREATS: MALWARE

"Cause the ceiling fell in
and the bottom fell out
I went into a spin and I started to shout,
I've been hit! This is it! This is *IT*? **I T** *it*!"

Orange Colored Sky
by Milton DeLugg & Willie Stein;
Best Rendition, Vocalist Jessica Neighbor with
Pianist Dan Zemelman

Nat King Cole popularized these song lyrics more than a half a century ago and small business owners, affected by "malware," are keeping the words alive today.

Malware, short for Malicious Software, is a general term for viruses, worms, trojans, spyware, rootkits and phishing attacks, inflicted by folks called "hackers".

As a business owner, you probably have already experienced some sort of cyber (in this case, through the internet) attack. At the onset of this destructive practice, the responsible parties were hacking for fame. The thrill of cracking passwords and poorly enabled security in general was what the protagonist sought.

> **J**onathan James: James gained notoriety when he became the first juvenile to be sent to prison for hacking. He was sentenced at 16 years old. In an anonymous PBS interview, he said, "I was just looking around, playing around. What was fun for me was a challenge to see what I could pull off."

Today the hackers are hacking for profit and it is _your_ money that they are seeking.

I am including brief definitions of the most common forms of malware. Although at first it seemed antithetical to my purpose for writing this book (simplifying), we all hear these words frequently so a little education on the common ones seems appropriate.

Virus

A small program that replicates itself over
and over, spreading copies to other
programs on the computer that the virus
has infected. At some point, the infected
computer gets so clogged up that it may
become unusable.

Trojan horses

This malware uses subterfuge to lure you
into opening it. Often, users click before
thinking to take advantage of an irresistible
free offer. A "backdoor" is created,
establishing easy access for the perpetrator
to come back later to wreak more havoc.

Interesting Factoid: The Trojan Horse is a
tale from the Trojan War. After a fruitless
10-year siege, the Greeks constructed a huge
wooden horse, and hid a select force of 30
men inside. The Greeks pretended to sail
away, and the Trojans pulled the Horse into
their city as a victory trophy. That night the
Greek force crept out of the Horse and
opened the gates for the rest of the Greek
army, which had sailed back under cover of
night. The Greek army entered and
destroyed the city of Troy, decisively ending
the war.

Rootkit

The malicious "planter" of what is called a rootkit, takes control over your computer system without you realizing it. They can then spy on your activities, modify your programs and gain access to all of your private information, for further malicious use. Rootkits are particularly difficult to detect and time consuming to eliminate. Rootkit-canals (sorry) are best left to a professional.

Worm

A worm is like a virus on steroids. It has the additional capability of being able to spread itself to other computers on a network.

A worm may carry a malicious program, such as a Trojan horse, which gives a hacker a backdoor entrance to your computer.

Phishing

Ever received this email?

> *"During our regular verification of accounts, we were unable to verify your current information. Please click here to update and verify your information. Failure to do so could result in a temporary hold on your account until it is updated."*

"Hah!" you say. "This ain't my first trip to the rodeo" so you immediately search for that magic button called "Unsubscribe." You have just been phished. Either way, unsubscribing OR updating your information, you have given some very bad guy (unisex) access to your computer.

If there is any one subject worthy of immediate frenzied attention by a business owner, THIS IS IT! If you are not following at least the basic best practices for network security, stop reading and go get protected!

Once installed, anti-virus software must be updated regularly (a process called virus definition updating) and your license renewed regularly.

Where is the Malware Risk?

All computers connected to the Internet are at risk to varying degrees. What you allow your employees to do with your business PCs defines what risks you face and their severity. These activities expose you to potential threats:

- Email and instant messaging
- Online transactions
- Web surfing and browsing
- Social networking, blogging and sharing media
- Online gaming (At work? But it happens so frequently!)
- Downloading music, videos or other recreational activities (these activities are costly in so many ways--using up bandwidth (the capacity or volume of your internet line) that should be available for business use is among the most costly).

You don't really need to know a good deal about the various forms of malware. You just need to know that *any one of them* can be extremely destructive to your business, potentially putting you *out of business* for hours or days and in some cases, permanently. Just as you would not leave your car unlocked at a shopping mall, you need to "lock" your business network as tightly as possible.

Useful Tool: An Acceptable Internet Usage Policy for your employees is a very helpful tool for improving your security. Many of our Clients ask new staff members to read and sign an agreement for acceptable use.

There are other good reasons for requiring this. Your internet lines cost you money. Most activities that you wouldn't want your staff to waste time doing, use up the bandwidth of the lines, making it unavailable to conduct your business.

A sample of an Acceptable Internet Usage Policy is included in the "Useful Templates" chapter of this book. There are many other versions, more and less detailed, available on the internet

Chapter 5

Spam, Spam, Spam, Spam, Spam, Spam, Spam, Spam,

If you are under 50 years old, google this:
You Tube - Monty Python - Spam

In Monty Python's diner, there was no breakfast allowed without Spam. The allegory to internet usage is dead on. Like it or not, every internet logon is dished up with a heaping serving of Spam.

> "That cost [of damages to businesses, government organizations, schools and other nonprofit organizations] is projected to be $338 billion by 2013. It's easy to understand the economic impact that unwanted email has on our lives. We can't work without it, and we don't want unwanted email in our way."
>
> from St. Bernard's Red Condor, an award winning provider of email security

What started out as a relatively innocent but annoying advertising technique has now turned into an astounding 80 to 95% of all email sent today (depending on the pundit of your choice). Roughly half of these are made up of Adult Content, Health, Personal Finance and Education/Training but in the last few years, political spam has been gaining ground rapidly.

Spam is now harder to spot, often embedded in links or attachments. <u>One ill-thought-out click</u>, a link or virus is planted and your personal financial information is now a half a world away, ready to be sold and resold to multiple unscrupulous crooks.

By the minute, malicious spammers (there are some benign ones) are inventing new and more enticing ruses to get us to open their emails. A popular scam today copies a legitimate and well-known company's "look and feel," using their logos and tag lines. The consumer sees a brand they trust and the damage is done with a single innocent click.

Classic Example: Unbelievably, people fall for this grammatically absurd scam often:

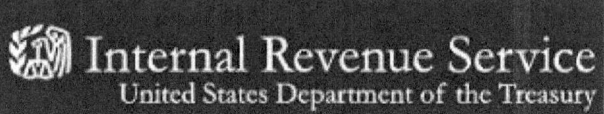

After the last annual calculations of your fiscal activity we have determined that you are eligible to receive a tax refund of **109.30**.

Please submit the tax refund request and allow us 6-9 days in order to process it.

You can apply for your refund online here.

A refund can be delayed for a variety of reasons. For example submitting invalid records or applying after the deadline.Please be carefull when entering your data.

Regards,Internal Revenue Service
© Copyright 2007, Internal Revenue Service U.S.A..

This particular scam uses the "fear" technique.

(Would that be "109.30 US Dollars", European Euros (= to 142.60 Us Dollars), Cuban Pesos (= to 0.05887 US Dollars), or Danish Krones (= to 19.1371 US Dollars)?

You no longer must contend with processed meat [sic] to use the internet. Today's tools for trapping spam before it hits your in-box are effective and inexpensive in comparison to the productivity loss your business would suffer without it.

TERYL L. BURT

Real World Story: A close friend of mine was out to raise money to fund his start-up company. Toward the end of the process, I received an email from him that said, "Click here to see me naked."

This spam is an example of a "joe job," defined by Wikipedia as "a spamming technique that sends out unsolicited e-mails using spoofed sender data. Early joe jobs aimed at tarnishing the reputation of the apparent sender but they are now typically used by commercial spammers to conceal the true origin of their messages."

This email went to thousands of people in my friend's Outlook address book, including his potential investors.

Being the well trained IT guy that I am, I didn't open it but I have been told that there really was a naked fellow in there.

(Incidentally, my friend got the money).

Chapter 6

I Invested a Small Fortune in a Tape Backup Device & Still Could Not Restore My Payroll

> A recent study discovered that, of companies experiencing a "major loss" of computer records, 43% never reopened, 51% closed within two years of the loss, and a mere 6% survived over the long-term. [1]
>
> [1] Cummings, Maeve; Haag, Stephen; and McCubbrey, Donald. 2003. *Management information systems for the information age.*

The typical small business owner has a few . . . ahem . . . details on their mind. Remembering to change the tape backup cartridge EVERY SINGLE DAY is a challenge. Taking it off-site, to someone's home or to a bank, to deposit in a safe deposit box, adds to the complexity. Replacing the media every six months—who knew? Whose job is that to remember? And the most common error my business sees, has been dubbed by our staff as "the air backup." When trying to restore a file or a document during a crisis, we often find that no data was recorded in the first place. Testing "restores" regularly is a best practice that is often overlooked.

It is never too late to panic (which is what you should do right now if you do not have "adequate" backup capabilities and have not tested a restore this month, quarter or year).

Reliable backup and restore of your company's data and documents in a timely fashion is crucial to the health of your business. As businesses feel increased pressure applied by business requirements such as business continuity, regulatory compliance and e-discovery (searching for electronic data to use as evidence in a legal case), it is vital to implement the right business-appropriate solutions.

"Adequate" is a key word because data storage needs just seem to grow and grow exponentially. Backing up all critical data at least daily is important; the ability to find and restore the lost or deleted data quickly, is paramount. This is particularly crucial for 24/7 operations where backing up can affect the performance of the server and potentially slow the users down.

> **A** **Beard Scratcher:** We get our teeth cleaned every six months, we renew our business insurance every year and we change the oil in our cars every 5,000 miles. These are preventative measures about which we don't think twice.
>
> Yet, safely backing up business data is not always given top priority, often because of the hassle, sometimes the cost.
>
> *It is not optional.* It may be more costly and bothersome than the basic necessities listed above but protecting your data is *far less expensive than the alternative.*

Considering the differing ideological approaches and the multitude of solutions on the market, selecting the correct backup and recovery solution for your business can be a mind-boggling issue. My company often suggests that you look at the most current technology first. Then back track if it does not seem appropriate to your environment, if the price is not right or it is just simply too bleeding edge for your sensibilities. The primary choices are described below.

The Holy Grail

A top notch solution for secure backup today is a **Managed Business Continuity Service.**

The best of breed includes a physical device that includes features such as:

1. **Complete application & data backup** - Backs up the programs as well as the company data.

2. **Granular restore options** - Enables recovering a single email, mailbox, file OR entire database.

3. **A standby server** - Built-in special software that allows the device to function as a standby server if a network file server fails. Once the real server is repaired, the information is transferred back to the server. If the server hardware has to be replaced, the device can restore the information to "dissimilar" hardware. (File server models are frequently retired (phased out) and replaced with updated models by the manufacturer so it is unlikely that you could purchase the exact same

hardware several years after your original purchase).

4. **Multi-site off-site replication** – The data is highly encrypted for security; it is copied to multiple locations around the country for added security (and to avoid issues with geographical natural disasters such as earth quakes, hurricanes, floods or tornados).

5. **Near Real-Time** – Backups can be scheduled for as often as every 15 minutes.

Everything is automated so there is no need to worry if someone remembered to change the backup tape, take it off-site for security or routinely test to make sure that there is data available to be restored. Part of the service package includes 24/7/365 monitoring so that the service provider is alerted to any error conditions and can work on correcting the problem before it grows.

These devices address multiple requirements of the backup, disaster recovery and business continuity needs of small and medium size business environments at an affordable price.

A Good Choice

A **disk based backup** strategy avoids all of the hassles of human touches and managing the physical "media" (tapes) required with a traditional tape backup. While a hardware tape backup unit with tape backup software used to be the budget conscious choice, the price of disk storage has declined substantially over the last few years.

The most compelling reason to consider disk based backup is performance. Backing up and restoring to disk is considerably faster. The hardware lasts longer and is not subject to environmental concerns like tapes are because they are in sealed cases.

A drawback to disk based backup is the lack of "data archiving" capability. This process allows you to retain multiple copies of your data. Businesses increasingly want data archiving for a variety of reasons, including government regulatory compliance. Also, with disk based backup, there is only one copy of your data available at any given time. If your network gets a virus and it is backed up before detection, there is no way to recover a good copy.

Additionally, if you keep your data all at one site, a disaster (such as a fire or flood) could wipe out everything.

An Option

On-Line backup services are an option for backup requirements that are not unusually large. If you have a lot of data to backup, it can still be viable but variables such as internet connection speeds become important considerations. (Your IT service provider can easily determine your precise data storage requirements and match that to the capabilities of an on-line backup service). Companies that have minimal IT staffs often find that online backup is a good option since daily human intervention is not required.

The drawbacks are the same as for disk-based backup, except that your single copy of your data is not on your premises, so can be retrieved after a physical disaster.

Getting Moldy

Tape based backups (TBUs) have been the solution of choice for many years. Today, reliability, cost and the hassle factor are causing businesses to consider more efficient options.

Companies are storing more and more data, archiving data requirements are becoming more stringent and the speed at which a TBU does its thing is becoming more and more inconvenient, particularly for businesses with lengthy business hours or overseas operations.

Real life Example: One of my Company's best Clients is a non-profit organization with operations in California (15 Users), Viet Nam and Cambodia (100 Users). When we are busy at work all day, the Viet Nam users are sawing logs, and vice versa. There is no time, day or night where it would be convenient for us to run a tape based backup because of the performance lag (and resultant productivity loss) that would be created on their network.

Keeping track of the media, (the actual tapes) can be a logistical nightmare. Tuesday's Week 1 tape gets mistaken for the Week 2 tape and gets overwritten so at least a full day's worth of data is unavailable. Tapes are fragile—they break, get dirty, the tape drive heads need cleaning. They need replacing at least twice a year and more often for larger companies. Why don't we use cassettes for music anymore?

Restoring data from tape creates another type of nightmare. While you could get lucky, it is something like finding a specific sentence or song on an old cassette tape-- fast forward-reverse-fast forward-reverse, ad nauseum.

The likelihood that the person who did the erasing or deleting can give you an accurate file name (precise, to the character) to look for the data is less than 50%.

If your business owns a fully functional tape backup, there is no need to rush out and replace it. As long as you regularly test restores to make sure the data is there, it probably makes sense to keep it through the end of its useful life (generally considered to be three to five years in a corporate environment).

TIP: Some companies, even data centers, combine disk and tape backup as their solution. Since backing up to disk is so much faster and causes less disruption on the network, this becomes the first step.

Once on disk, the data is backed up again to a tape backup where the tapes can be labeled and archived.

Nothing trumps a managed business continuity service for reliability and ease of use.

Chapter 7

My Insurance Agent Wants to See My BCP?!?

Do not blush! Write up a "**B**usiness **C**ontinuity **P**lan" and ask for a discount on your business insurance!

A Business Continuity Plan (BCP) is a comprehensive strategy that plans for natural disasters as well as common events that could have a significant impact on business operations. It collates many individual elements, (including those that are vital to a Disaster Recovery Plan (DRP)) into one big ginormous, comprehensive plan that touches *all* functions of a business (not *just* technology).

A Disaster Recovery Plan (DR) focuses on IT. Loss of data could mean losing patient or Client files, accounting data, critical company records, legal records, email trails, outstanding orders and a host of other important information.

Most companies cannot afford to lose their data and keeping data safe at all times is a critical challenge. Yet writing a sound plan is often neglected because of lack of expertise, lack of staff to create and maintain this information or commonly today, lack of resources. It is also sometimes due to failed memory (ours).

BCP planning goes well beyond DR because it encompasses all aspects of company operations.

Small business owners often say, "I'm too small to need all this fancy paperwork." The truth is, everyone needs a BCP. How complex the plan needs to be is up to the business owner and pressure, if any, from auditors and external parties such as regulatory bodies, strategic partners and even customers.

The plan could be a simple document (paragraph?) about who is in charge in the event of the untimely death of the owner? Is there a hard copy of Accounts Receivable? Inventory? Where are they kept? Who has keys to the building? How do we evacuate the staff? How do we contact our insurance company? Who *IS* our insurance company?

Business Continuity 101: Write (at least) the basics down and store the document safely!

Elements to a Basic BCP

Described below are some basic documents that businesses and organizations should create to protect themselves, all of which are necessary components of an effective BCP.

1. Written Network Documentation
2. Technical Diagrams (Should be included in the Documentation set above)
3. Compliance Policy (as required)
4. Summary Tech Info Sheet

Written Network Documentation Set

A network and its associated software and hardware are significant business investments. Maintaining accurate documentation of the design and individual elements (in sum, called "infrastructure") is critical to the success of ongoing operations. A thorough documentation set includes information about your network necessary for tracking ongoing technical work, disaster recovery and future strategic planning.

The objective of "living" network documentation is to keep the information you need for problem resolution readily available. All changes and upgrades made to your system should be recorded here and this manual should be kept with other operations manuals for your business.

A documentation manual should be presented in a format that can be used by any current or future network consultant who may need to work on your infrastructure.

Technical Diagrams

A network diagram is essentially the same as a blueprint for a building and is just as important as the written network documentation set (and should be stored within the document). It is an essential element to making things functional, secure and safe. Network diagrams should be created following industry best practices and standards in IT, and should incorporate the laws and regulations mandated by building departments, landlords, utility companies and any other compliance requirement to which your location is subject (state or local for example).

A network diagram is also a key DR tool. The pressure for excellent customer service is growing rapidly. It behooves the business owner to create and maintain a solution that allows for efficiently conducting business as usual, no matter what situation arises.

With an accurate, up-to-date diagram at hand, any qualified IT person will be able to quickly review it and understand how the network is configured. This will make a major difference in your "time to recovery."

A **Practical Example of what is called Workforce Continuity Planning:** If you live in the San Francisco Bay Area, you no doubt vividly recall the Loma Prieta earthquake that occurred in 1989. TSG is located in the "east bay" and our staff members came from all over, many from the north and south bays. The collapse of a portion of the SF Bay Bridge blocked many of them from getting to work in the following days— year, actually.

Our CFO developed the "hopscotch" plan. The person living the farthest away drove to the next farthest person's house. They then went in the 2nd person's car to the next farthest, and took the 3rd person's car, and so on, and so on. The wear-and-tear on our staff of extra driving was reduced and the costs of the extended commute were shared among a group.

Because earthquakes are so prevalent in this area, this practice is now a permanent part of our DRP.

Summary Tech Info Sheet

A "Tech Info Sheet" is a homegrown document developed by our senior network engineers, although most IT service providers have their own version. It is merely a snapshot of critical information about a Client's network. (All of this information should be documented elsewhere in detail). This is just a "ready reference" with information you might need in a hurry (such as phone numbers for Tech support, software versions, etc.).

Compliance – special considerations for specific industries

Different types of businesses are subject to specialized government mandated compliance rules. These regulations require "proof of implementation." To be compliant, organizations must ensure that all employees are trained and understand requirements for SOX, HIPAA/HITECH, FERPA, OSHA, FMLA and other state and federal regulations.

You are probably already aware if you are subject to special compliance rules. What you may need to do is make sure your employees are on board. This can be handled by having them sign off after reading required documents when they are hired and revisiting the documents in short

company meetings as often as is required
by law.

Check This Out: There are public
domain templates available that make
a good starting point for you to create
these tools yourself (i.e., MS WORD
Templates).

If you work with an outside support group,
they can provide you with more detailed
and specific templates.

Ask for a quote for having them do the
work. If it is affordable, they will likely
catch more detail and create a more
thorough document.

Chapter 8

OK, I Probably Need Some Outside Support

How often have you had a problem or issue come up and no one on your staff remembers how to configure a SPAM filter or the password for the network components? During the 9/11 emergencies, companies that had done outstanding jobs on disaster preparedness were shut down because they didn't have correct server and router passwords.

A good outside support group should guarantee to have staff that knows and keeps track of your site (and all of your critical information). They will keep the written documentation, diagrams of your environment and your Business Continuity and Disaster Recovery plans up-to-date. Your staff may change but you can be sure that someone maintains current skill sets on your technology.

> **R**evelation: **Experience Trumps Certifications.** I frequently see ads seeking IT people that start by listing scads of "Required Certifications."
>
> In reality, it is not that difficult to pass the Cisco, Microsoft or other exams, so "certified" does not *necessarily* translate into "able." (One time when we were in a bind on a job that we really wanted, our receptionist crammed and passed the missing Cisco exam that we needed).
>
> While I am in no way diminishing the value of certifications, real-world experience is a very good indicator of quality technical capabilities.

Some advantages of bringing in outside support, permanently or periodically, include professional advice and assistance in the following areas:

Strategic Planning

Technology consultants can help you write technology plans that work. IT service providers spend a great deal of time upgrading their skills and staying current with technology trends and "best practices" (considered by leading industry professionals as a whole to be the most efficient and effective way to do things). A good strong vision, a year or more into the future, will help you make wise technology choices today.

Documentation

A competent outside support group can help businesses identify and document those areas that are critical to the operations of the company.

Having an outside person work on updating network documentation often helps the organization tighten up their operations and fix potential problem areas before they erupt into major disruptive issues.

Fresh Perspectives

An outside party can review your environment with "fresh eyes" to help spot a developing problem or educate the business owner about new and/or more efficient options. Third-party support groups must keep their knowledge and skill sets current to remain competitive.

Much of the emergency work that requires calling in an outside support group on an "incident" basis (remember, generally the most expensive) is after a knowledgeable Network Administrator attempts an upgrade out of their current skill set and runs into a problem.

A good example of this is a network file server operating system upgrade. These are generally only completed every three (3) to five (5) years and there may have been several versions of the software released in between times. File server operating system upgrades are a complex undertaking and internal staff may not have experience installing or using the new version. They could take a class, attend a webinar or read a book, but hands-on experience is the most valuable tool for ensuring a successful outcome.

If the upgrade has to be corrected, finished or completely redone by an outside group, chances are it will be at a premium cost (particularly if it is an emergency "after hours" job). A qualified outside support group will be aware of the "gotchas" that can save businesses time and money.

Finding that "trusted advisor" everyone is talking about

It is a little like a good marriage--you need to work on it to make it successful. Whether you choose to make that advisor part of your internal organization or choose to keep them at arm's length, routine communications is vital to the health of your technology environment. Your advisor needs to be on the lookout for areas that will increase your Return on Investment (ROI). Scheduling regular meetings means they know what you are thinking about, where your organization is going and can articulate how technology can help you get there.

Whether you employ a full IT staff or outsource all or part of IT management, put in the legwork to find an outside support group that provides a full range of scalable service products to complement your needs. They must be objective and aware of all the issues facing each stakeholder in the business.

> **C**an't Stress Strongly Enough: It is vital for the business owner to participate in the selection of the outside support group and in the planning process.
>
> Too often, this is relegated to an admin person or in-house IT person, who may not be fully aware of the owner's "big picture" or may be threatened by bringing in outside advisors.

If you are going to pay them, trust them

When I was growing up with my siblings in rural Illinois, coffee was perking all day, every day and it was the great communication's facilitator. Any time of day or night, whoever stopped by was handed a cup of perked coffee, whether they asked for it or not. (There were a few highballs but that was not the norm).

Whenever we had to call a repair person, however, our mother hissed on his/her way in "Don't give them any coffee until they are done!" She had an inherent mistrust of service people. She was as friendly and kindly as could be—once they were off the clock.

This is often true in the technology arena. Small business owners are fearful of being ripped off by unprofessional, unknowledgeable or limited scope "consultants," probably for a good reason. There are bad to mediocre ones everywhere in a field that changes at breathtaking speed.

If you have done your legwork selecting an outside support group well, you have chosen someone with broad technical expertise in addition to the appropriate communications capabilities that allow you to "hear" what they are saying. They will be working proactively to improve your business. Their advice is based on far more experience than you probably have in technology. They understand your interests and will put your needs ahead of theirs while in your employ.

Ultimately, whatever happens is still all your decision.

Some time ago we worked with a business owner who we sensed would be difficult but we liked the nature of the business (fancy foods—yum!). As we were doing our discovery and documentation, to get familiar with the environment, we made recommendations about their existing equipment. Many of their network

workstations were old and needed frequent
repair. We wrote up a standard replacement
policy based on a 1/3, 1/3, 1/3 three year
replacement schedule to spread out the expense.

Convinced that we were trying to sell him a
bunch of new hardware (at about $4.79 profit
each, so not much motivation on our part), he
chose instead to continue to spend $300-$400
annually to repair *each of them*. He ended up with
crummy old workstations and a lot less money.
We referred him to a Geek working out of his
house and went off to find a Client that would
listen to the advice for which he was paying.

Budget TIP: The cost of replacing
legacy (old) equipment should factor
into the cost of not doing so.

Poorly functioning equipment leads to more
downtime. Productivity loss, elevated
employee stress levels, a decrease in
competitiveness, missed deadlines, lost
data, cash flow drain and poor customer
service resulting from downtime can never
be reclaimed.

If you trust them, pay them

Having to question a professional services bill is sometimes necessary but should not be the norm.

Good Habit: When my company is in the final stages of negotiating a new contract, we REQUIRE our staff to sit down with the Client and review in excruciating detail two sections of our standard Agreement:

    1) "What is Not Covered"

    2) "Client Responsibilities"

The Client already knows what *is* covered and in their enthusiasm about the relief soon to be had when the new support program starts, they often do not "hear" what is not covered. This can lead to billing disputes that are not pleasant for either side.

Nobody likes surprises (unless a large gem stone or a yacht is included).

Have a frank and clear discussion with your service provider in advance about what activities are billable and how billing is calculated. Then plan for some level of "unexpecteds". Service providers can predict how technology should behave but not how people are going to use it.

Microscopic examinations of every bill or expense, or regular cynicism at recommendations means that you have hired someone whom you don't trust and you are back to the task of managing IT instead of growing your business.

Chapter 9

50 Ways to Leave Your Current IT Support Vendor

Well, not really 50.

This may seem like an unnecessary point to discuss but it is important. No one wants to fire Cousin Leonard or the nice people who give you the Harry and David's Tower of Chocolate Cherries & Nuts for the holidays.

That is not a good reason to put off making an initially uncomfortable change. Many business owners delay the decision to replace ineffective technology support vendors far too long because they dread the confrontation or are not sure what to do instead.

Switching service providers can be expensive time-wise (i.e. learning curve) so the decision should be thought through carefully and based on real particulars. It is hard to let go of someone or some group with whom you have worked for many years. Quite likely, they are not all bad.

Unfortunately, there are certain things your business needs and you should expect. In

TERYL L. BURT

addition to standard maintenance of, and information about your network, some of the other things you should expect to be routine are:

- Regular communications, via the method of your choice (email, phone call, lunch) to check on new needs or requirements;

- Written reports about your network functionality;

- Notice of changes in the law that affect your business;

- Opportunities for getting stimulus funds or other government backed programs;

- Alerts about the release of a particularly virulent virus or other malware;

- Maintenance renewal notices (timely) for software, network devices, extended warranties, etc.;

- Regular big picture face-to-face reviews.

Your prospective new consultants can help advise you on what you are missing so that you have the ammunition you need if you choose to cancel services. They should be willing to sit in with you or even handle it if you prefer.

Be wary of any consultant that charges in like a white knight, to criticize and malign your old support group. A professional services provider will stay focused on how to make improvements for the future instead of re-hashing the past.

> **N**o need to be coy, Roy. It is your business, your livelihood, your children's education at stake.
>
> Times change. Needs change. Make the changes when the time is right, which is probably when you first see a problem developing into a pattern.

If you truly want to keep them, for whatever reasonable or unreasonable reason, tell them clearly and directly and in writing what you expect. If they do not perform satisfactorily after that, any guilt you were feeling will be assuaged. An ethical company will assist in transitioning to the new group in any (reasonable) way they can.

CHAPTER 10

IMAGINE . . . YOU MAY SAY I'M A DREAMER

Imagine . . .

. . . the day when you have almost no <u>Capital Expense</u>, no need to buy and maintain expensive equipment that you don't quite understand and that becomes obsolete before the ink is dry on your first brochure.

. . . <u>not</u> dedicating <u>anyone</u> on your payroll to manage an ever-more complicated mass of equipment or wasting your best sales person's time to fix Arthur's workstation because that sales person once un-jammed the printer and got dubbed "the IT guy."

. . . <u>knowing month-in</u> and <u>month-out</u> what your IT related Operating Expense will be and that you will have no whopping and unexpected technology repair bill in February, your worst month of the year, because your network file server crashed.

. . . that <u>your data is backed up, all day, every day</u>, and exists in several locations around the country with the highest possible level of security and the fact that Clara forgot to change the tape—*again*—is immaterial.

. . . knowing that <u>100% of your equipment is state-of-the-art</u>, the most current software versions, the best virus protection and the strongest SPAM filter available, with access to your system from anywhere there is internet access, any time, and as close to 100% uptime as anyone could claim. And . . . from any device. If it boots up, it's usable (yes, even if it's 10 years old).

. . . getting a <u>discount on your business insurance</u> because you have a written Business Continuity Plan (BCP) and Disaster Recovery Plan (DR).

. . . <u>getting whatever resources you need</u>, on a moment's notice, to add the 6 new staff members when your ship comes in in April, and also on a moment's notice, <u>eliminating that expense</u> when your ship sails out in August. You didn't spend thousands of dollars for software licenses that now sit idle. This is called "scalability."

Whether a business employs internal IT staff or uses an outside support group, the bulk of the IT staff's time is spent primarily in reactive mode, creating time constraints that leave the company vulnerable in a number of areas. This management method creates hidden costs because the dollars are spent in small increments--over time, usually the most expensive practice.

Consider the cost of bringing the technology environment current versus the revenue opportunity created if IT could look "upward," coordinating strategically with management to generate new growth opportunities. The visionary position could be internal or directed by an experienced outside professional IT team.

Personally, I *am* a dreamer but not about the possibilities described above. They are all <u>real</u> and <u>happening</u> <u>today</u> for successful growing businesses around the world.

The next three chapters describe the possibilities in more detail. They are probably *<u>more within your reach than you ever imagined</u>*.

Chapter 11

Managed Services:

The *PRO*-Active Way to Reduce *RE*-Active *Avoidable* Expenses

Living next-door to a firefighter is one of the most comforting joys of living in my home. Rune is always ready. Whatever the emergency, he and the rest of his squad is always ready to react within seconds. I am happy to pay my share of the taxes that goes to the fire fighters because I always know that 24x7x365, someone is trained and ready if my wooden home is threatened.

It could still burn down, though, if the emergency is too large for the available resources to handle in the amount of time available to "react."

In the IT world, emergency responders will likely always be necessary but the most important tip a fire fighter will give you is:

> *"Clear the brush away from your home ahead of time to **prevent** the disaster from occurring in the first place."*

Reaction, in IT emergencies is expensive and 24 hour on-call support has historically only been available to the big guys who can afford "squads" of on-call staff.

Managed Services (MS) is a broad term used to describe proactive, ongoing management, remote support and planning of IT environments on behalf of a business or organization. Services often include support for your users via a "HelpDesk" and this support is often unlimited. Generally, these services are provided at a flat monthly fee.

As I briefly described in Chapter 3, managed service providers are banking on the fact that preventative maintenance will minimize disasters. They can offer these flat fee services at much lower rates than they can offer either incident or emergency response.

Depending upon your specific needs, choosing coverage is like choosing from a Chinese restaurant menu. There are all kinds of min-and-match coverage plans for elements such as:

- File Server(s)
- Workstation(s)
- Network Components (switches, routers, etc.)
- Security (Firewalls and other layers of protection)
- Disaster Recovery and Business Continuity Services
- Backup
- HelpDesk
- Software and software patching
- Point of Sale equipment (like cash drawers)
- Custom Requirements

Fully managing your environment maximizes the use of your resources for greater productivity and provides early detection of potential issues that can be resolved before they escalate.

Another Tip Worth Serious **Consideration:** Managed Services (MS) is a flexible solution that is worth investigating or expanding upon if you are already using some services (i.e., payroll).

It can mean access to the most worry free, innovative technical solutions available, no matter what the business size.

Greater Automation = Greater Efficiency

Chapter 12

1-800-The-Cloud

Incomparable Performance
(And You May Be Pleasantly Surprised at the Cost)

Ask 5 people what The Cloud is and you will likely get 5 unique answers.

Some people will say, "It's putting your stuff in a data center" instead of keeping and maintaining it on your premises.

Some people will say it's "virtualization," the latest and greatest process of separating software from hardware, thus getting more with less.

Most people will answer in a way that furthers their particular cause. Someone who sells security will spin it as the best possible security solution. Someone else who sells backup will tell you it's a way to achieve absolute safety of your data.

TERYL L. BURT

The reality is, those are just tools. A _data center_ is just an extremely safe _building_ with a bunch of top-notch equipment and phone lines; expert IT engineers install and maintain the equipment on behalf of businesses.

Virtualization is just a _technology_—a cool and innovative one—that allows us to manage our resources better. Most individual computers use only a fraction of their power at any given time. When you virtualize, you put a bunch of computers "virtually" (by way of special software) into one piece of hardware, maximizing all of the resources. Less equipment, less power, less cost.

Very simply stated, moving to The Cloud allows the business owner to eliminate purchasing, configuring and maintaining complex and expensive hardware and software. The Cloud vendor takes responsibility for _everything_.

I recently attended a conference on Cloud Computing. To paraphrase Kent Erickson, CEO of Pointivity and a well-respected, highly successful early adopter of taking his Clients into The Cloud,

> "Cloud is a way to deliver business outcomes to the business owner that is utility priced (meaning you pay for only what you use, as you use it, that month) and that shifts the risk for designing and maintaining the appropriate system to the Service Provider. If something goes wrong, it is on our dime. We cannot charge our Clients any more money and that is guaranteed in the agreement."

What you get is a total solution, with no surprises. Instead of buying things piece-meal, you receive a completely integrated "whole" that is delivered to the desktop of your choice.

Cloud Integrators are in the risk arbitrage business. They are betting that they can do it better, faster and at a lower price point than most businesses could even begin to realize. Moving your business to The Cloud (partially, in a hybrid solution or entirely) could be the safest, most cost-effective and rewarding business decision you could make today.

The Benefits

- Well Defined Service Level Agreements (SLAs)
- Superior Performance
- Lower Cost
- **Security**
- Scalability

Well Defined Service Level Agreements (SLAs)

When you engage with a Cloud Computing vendor—my company calls itself a *Cloud Integrator*—you and the vendor agree upon an SLA. The commitments in these agreements are spelled out to the *nth* detail so <u>you know exactly what is covered and what your costs will be</u>. They also spell out your recourse should anything not be delivered as promised.

Superior Performance

Performance guarantees are outlined in the SLA. Because of economies of scale, Cloud Computing providers offer resources and levels of service that are vastly superior to anything a small business owner could afford to purchase and maintain themselves.

For example, TSG's data center provides high-speed lines from seven (7) major carriers, and each one has the capacity to carry the *entire data center* (thousands of businesses). If AT&T has a service interruption in our data center region, our Clients are seamlessly moved over to XXX or YYY. Most Cloud Computing providers provide at least a few carriers but not necessarily enough to handle everyone.

We recommend that businesses install (in their office) a secondary backup line from another vendor so that if there is a local outage by a major carrier, the business can flip over to the other line. It can be a less expensive, slower speed line with less capacity, since it will only be used in emergencies.

Watch **Your Step**: Moving into The Cloud, you get 24x7x365 support. Be cautious about claims of "100% Up-time Guarantees."

Chances are, you will probably experience 100% uptime but: *things do go wrong in technology.* The odds of destructive events happening are far fewer and the response is far quicker in the hands of experts.

Lower Cost

> *"Cloud and hosted services have disproportionately high benefits for small companies because the cost of in-house IT, support, hardware and the like really hurts their operating expenses and bottom line."*
>
> –Paul DeGroot, Analyst, Directions on Microsoft

Moving a business into The Cloud means you "pay as you go." This is "utility based pricing."

A business or organization can grow or shrink based on seasonality or other factors, without making a capital investment in resources (like software licenses) that will sit idle for the rest of the year. You avoid the Capital Expense required to purchase network file servers and most of the requisite network components.

Security

Security remains one of the major concerns about Cloud Computing. The bad stuff hits the news and the advances are sometimes too complicated to understand.

For the small to medium sized business owner seeking to offload the expense and hassle of purchasing and maintaining an in-house network, choosing a reputable, experienced Cloud vendor provides an increasingly safe and secure environment.

Scalability

Scalability simply means that your business can grow and/or shrink on a moment's notice. Peaks and valleys are no longer the hassle and cost they used to be. A business can add users, additional memory to increase speed, more storage space as data grows, and even another file server if there is a large change in the number of users, at a moment's notice.

Likewise, all of those things can be reduced in minutes. You pay only for what you use.

Chapter 13

Cloudy Thinking

Is "The Cloud" Right for My Business?

Not everybody is ready to embrace Cloud Computing. Not everyone can get their arms around the concepts yet. Some business owners can't conceive of operating their business without a harried IT person or technician physically present, running around on-site to deal with all of the issues that come up all day long.

Managing technology is generally no fun for the business owner. It is a hundred times worse than having to take the garbage out, straightening the conference rooms chairs all day long, calculating sales tax returns, getting billing out on time, etc.—all combined. Since it is unavoidable, you have to learn to manage it or pay to have it done.

The business owners who are ready to embrace The Cloud are the ones who want to focus on growing and expanding their businesses. They want to focus on doing what they set out to do before they found out that to manage information these days you must employ technology and technology is complicated, expensive and short-lived.

The business owners who are ready are the ones who want to shift the risk to an expert who lives and breathes technology and is good at it. They want to spend their precious cash on business ideas and growth.

ISN'T "THE CLOUD" EXPENSIVE?

Quantify the Total Cost of Ownership (TCO) of an internal network before deciding for yourself.

Add up the cost of the hardware and software, the add-on licenses, making sure to include the annual maintenance cost for things like tape backup software and a firewall. Refresh all of this on a 3-5 year cycle. That's all Capital Expense (CapEx) to pay for fixed assets that are often depreciable, but not tax-deductible.

Then add in the payroll cost for an internal IT person with benefits at 20-25%. Don't forget to add in the cost of outside consultants because you will undoubtedly need them at some point.

The proof of "expensive" is in the TCO calculation. In most cases, what you have in-house is a homegrown network, with a medium to high degree of risk because your security is nowhere near what it should be.

Is "The Cloud" Risky?

With any new computing development there come risks. Even today, mature technology platforms carry with them considerable risks. The Cloud helps to reduce some existing risks while creating some new ones. The most obvious risks in a move to The Cloud are:

- Maturity of the provider
- Connectivity (internet)
- Security

Maturity of the Provider

In The Cloud, the service provider holds almost all of the risk, which should be highly attractive to business owners.

They must take full responsibility for the service that they deliver. Thus, the big issue becomes one of trust. The business owner needs to trust the service provider enough to understand the magnitude of the risk being shifted and believe that they will be ultimately accountable.

For example, if a business purchases Microsoft Exchange Server software to manage their email in-house, Microsoft's license agreement makes it very clear that the buyer is responsible if anything goes wrong. In The Cloud, the service provider is responsible since the Client is "renting" the licenses from the provider, on a pay-as-you-go monthly basis (utility based pricing).

Choosing an experienced, knowledgeable Cloud service provider is the most important homework in a business owner's backpack.

Connectivity

Internet connectivity is at the crux of a Cloud Computing solution. If internet access is interrupted, everything is down. Much, and in some cases, all of the company's work is interrupted.

All true—is this much different from working in-house? Yes, because Cloud Computing providers offer multiple access lines to the internet.

Security

As far as risk goes, the most insecure place for your critical data is probably at your business location. The likelihood of something happening to a data center is infinitesimal compared to the chances of something happening at a business location.

Data Centers do not burn down, it is very difficult for a person to physically even *enter* data centers and it would take a backhoe to trench around the entire building, without being seen, before someone could take out the phone lines.

CHAPTER 14

SECRETS

As I started thinking about secrets that might be helpful to a business owner when choosing an IT vendor, the idea that good results really is mostly about finding a *trust* relationship became even clearer.

Here are a few thoughts I can share based on my thirty two years in this business. (There are good guys and bad guys everywhere so this assumes you picked a good IT vendor).

1. Your business agenda should drive the technology agenda, not vice versa. This is on you, to make sure it happens.

2. When the business owner gets involved in the engagement, a better outcome is assured. Yes, it takes time; it is easier to get to the real facts if the main stakeholder shares real information, untainted with fears of job security or inner-office hullabaloo.

3. We (IT Consultants and Service Providers) do not book extra hours. In fact, we agonize over every billable hour that is beyond what someone (us or you) thinks is how long something "should take." Given the lightning fast pace of technology, no company can know everything. What we get paid for is years of <u>education and experience</u> to know the <u>fastest way to figure things out</u>.

4. Small businesses often have <u>*unique-er* problems</u> than large businesses. These are things that are not documented by Intel or Cisco or Microsoft so we cannot just look them up. At the risk of sounding arrogant, chances are <u>we can figure it out faster than you can</u>.

5. Trustworthy IT support vendors <u>want to help your business succeed</u>. If you grow, so do your support requirements.

6. <u>Preventative measures</u> really do <u>make a difference</u>. It is just like changing the oil in your car.

7. Blackberries are a pain. Expect that <u>configuring Blackberries</u> onto your network <u>will take time</u>. Most other cell phones are more straightforward. (I am not dissing Blackberries, just stating fact based on my company's experience to date).

8. Smaller *can* be better. Consider your personal style when choosing an IT group. A small company will likely give you <u>more personalized service</u> because you are more important to them.

9. The <u>most common reason for dissatisfaction</u> with results is <u>misaligned expectations</u>. Spend the time it takes to make sure your vendor understands what you want. An ethical vendor will say "No" if they do not have the capability to produce the results you want or don't agree with your decision. We advise, you decide.

10. You <u>get the most</u> out of your service provider <u>by contracting</u> with them in some way rather than depending upon incident (time and materials) support (remember, the most expensive, least satisfying way to do business).

I DON'T SPEAK GEEK

CHAPTER 15

HOW DO I KNOW YOU ARE TELLING IT LIKE IT IS?

Although Mark Twain once said, "Few things are harder to put up with than the annoyance of a good example," nothing, really *NOTHING* beats true success stories. Hearing about someone else's experience can help a business owner decide on a technology strategy to fit the unique needs of their business.

The Case studies below include Clients with whom my company has worked.

The testimonials following, are past and present commentary on how my company, *TSG,* conducts business.

Case Studies

Example: Cloud Computing for Chain of Retail Stores

The Company

Barons Jewelers is a family owned and operated retail jewelry chain that has been serving the San Francisco East Bay area since 1967. The staff at Barons is committed to providing the finest high quality jewelry, from some of the most famous designers and watch manufacturers *at affordable prices.*

The Situation

Baron's absolute commitment to exceptional service, quality and affordable prices has served them well, and over the years, they expanded from a single store into a multiple store operation, located in major shopping malls. While 25 years ago Bart Heller, Barons beloved founder, succeeded handily by conducting much of his business with a smile and a handshake, his son Ronnie recognized the evolving need for efficiency, communications, speed and above all, security. Ronnie knew he had to invest in greater technology to continue to drive his company's growth and protect his staff and clientele.

The Challenge

Barons was running their network operations on an old and increasingly unreliable VAX system. Their vertical applications software vendor was no longer supporting the operating system. Changing software was not an option: as a busy seven-day a week retail store chain, they have mandatory requirements such as point-of-sale, bar code, communications between the stores and government mandated customer information security issues.

The Choice

In June of 2007, Ronnie met one of TSG's owners at a professional event. Ronnie was immediately intrigued with the notion of "outsourcing" his technology and the inherent headaches managing IT entails. An outgoing, friendly and ambitious business owner, he wants to work with his clients, not figure out why the backup didn't run again last night or whether he has enough licenses to add a user.

The Response

Barons time-line was short—they needed a solution quickly as the old systems were failing more and more frequently. They needed iron-clad security for credit card processing and client information. And because such a large over-haul was required, they needed a cost-effective solution that did not require a huge capital outlay at one time, on top of the major cost of upgrading their main application.

The Result

In September of 2007, TSG "on-boarded" Barons into their state-of-the-art data center, providing all brand new, scalable high-end equipment, all of which runs transparent to the company. Now they connect through the internet via Citrix, the premiere virtualization and remote access software for delivering applications over a network and the Internet. Citrix allows users to connect from anywhere there is internet access, whether it be from home, while traveling, or an international internet cafe.

Instead of purchasing PCs when the existing point-of-sale and accounting workstations need refreshing, Barons now buys "thin Client" workstations, for about half the cost of a PC. Thin clients have an optimized operating system and graphical user interface with no applications installed locally. This server-based computing alternative requires nearly zero administration at the desktop, which significantly reduces the cost of managing IT infrastructure.

Sure, things happen. The internet goes down, cabling needs upgrading, weather knocks out phone lines--these are now momentary bleeps for Barons instead of constant daily worries. In these turbulent economic times, Barons has assured itself the best possible opportunity to continue to grow and prosper with confidence in their processes, fixed monthly costs, current and up-to-date licensing, and top-notch security.

Best of all, Barons now has an owner who is free to grow his business.

Example: Cloud Computing for Non-Profit Organization

Case Study: East Meets West Foundation

The Organization

The East Meets
West
Foundation
(EMW) is "the
foundation for
learning,
healing and
health."

EMW's work reflects the belief that every person deserves access to clean water, proper medical treatment and a solid education. Without these fundamental elements of a good life, children cannot thrive and adults cannot be fully productive members of society.

Founded in 1988, EMW has a 22-year track record of innovative and effective work in Vietnam and a vast portfolio of completed projects. EMW projects and programs are known for their high quality, long-term sustainability, emphasis on results and significant scale.

The Situation Today

Many non-profits face the challenge of having limited resources for technology, in a complex world where automation is crucial to survival. The mission, of course, is to fund the programs. Frequently, whatever technology exists consists of a hodge-podge of ill-suited, donated, legacy parts that over the course of time, cost more to keep operational than new equipment might have at the onset. Generally, someone within the organization is appointed to be the "IT staff" but given no budget and no time to create a vision for the future.

The Challenge

With three offices in Viet Nam, and one in California, the frustration of working with non-integrated software, a network that went up and down daily and the inability of a committed staff to work remotely, even though they were willing to devote personal time in support of their mission, created a near melt-down for the East Meets West staff in late 2005. Keeping the accounting data synched between the offices was a major headache.

The Choice

When East Meets West (EMW) decided to seek outside professional services, they took the very logical approach of asking for referrals from some of their other vendors, colleagues and competitors. TSG Networks was named multiple times as an experienced professional IT services company, savvy in the special needs of non-profits.

TSG met with the EMW management team and listened carefully to the mandatory, the logical and even the far-fetched "wish list" goals EMW wanted to achieve. Drawing upon almost three decades of experience, TSG translated what was learned in those conversations into a three-year technology plan that respected EMW's budget while outlining an encouraging growth path.

The Response

After performing a detailed audit of the existing equipment, it was clear that the capital outlay for replacing required equipment would be cost-prohibitive. Key elements, such as a firewall, were non-existent and the file server was a re-worked workstation class pc that was over 6 years old. Without a qualified in-house IT person, the costs for maintaining this environment were just not logical.

The Result

In September of 2005, TSG "on-boarded" EMW, providing brand new, state-of-the-art HP file servers housed in TSG's network operation center (NOC). By employing Citrix software, the premiere virtualization and remote access software for delivering applications over a network and the Internet, EMW not only had stability, but scalability and 100% maintenance of their file servers at a fixed, budgetable fee, which amounted to a fraction of what they were paying to repeatedly band-aid the old equipment. For the first time, the accounting files were being updated in real time from the overseas offices. The staff was most excited that they were now able to access their servers from home, on vacation—virtually anywhere that Internet is available.

EMW opted to combine TSG Desktop Assist with their hosting program. With unlimited helpdesk support via either telephone or email, users can get immediate response to just about any issue that comes up. Better still, more than 95% of the time TSG can resolve the issue remotely.

Twenty-four months later, EMW went one step further and worked with TSG to move their environment into TSG's state-of-the-art data center. EMW now operates as close to 100% up-time as one can get. Their costs for these services are equivalent to about half the cost of an in-house IT person. What were once far-fetched wish list items for EMW are now in process.

Example: Project Outsourcing for Private Property Management Company

Case Study: K & S Real Estate Management

The Company

 **REAL ESTATE MANAGEMENT**

Established in 1958, K & S Company Inc. ("K & S") is a full-service property management company that provides services to owners of shopping centers, office buildings and apartment complexes. K & S has a complete staff of property managers, accountants and leasing agents who handle half a billion dollars in real estate assets. The average employee has worked at K & S for fifteen years and the average client has relied upon the company for eighteen years.

The Situation

While K&S had an impressive half decade track record, the volatility in the real estate market led K&S's new Chair and CEO to seek professional guidance for their technology environment. In order to upgrade to a more state-of-the-art vertical application, the underlying infrastructure had to be brought up to today's standards.

The Challenge

The operating system underlying K&S's network, Xenix, was no longer supported, and there are very few networks specialists left who are familiar enough to work with it, even in the technology-rich San Francisco bay area. The hardware was failing and disaster seemed imminent. In-house IT capabilities were limited to knowledge of the existing system and little documentation existed to allow anyone else to assist. Losing data would have been catastrophic to the Company.

The Choice

In June of 2006, TSG was introduced to K&S management by a friendly competitor. TSG quickly assessed the big picture, and provided several possible solutions to K&S. Because TSG takes a 360 degree view of every Client, we were a logical choice to address all areas of concern to management. Security was at the top of the list because they deal with so much confidential information.

The Response

TSG designed a completely new environment for K&S, including new cabling, network file servers, all network peripherals and workstations, high-speed phone lines and appropriate licensing for all applications software. We brought in an outside contractor with Xenix experience to help through the tedious conversion process and arranged a competitive leasing package so that everything could be handled at once. Because weekdays at K&S are so busy, TSG arranged after hours and weekend time to complete the work without interfering in K&S's day-to-day business.

The Result

Today K&S is completely converted to a state-of-the-art environment which is fueling their growth by providing top-notch efficiency and security. Their savings in annual IT costs covered the cost of the network upgrade. TSG continues to provide technology services to augment a part-time in-house IT person.

TESTIMONIALS
OLDIES BUT GOODIES

"Name me another support organization that will go into their office on Thanksgiving Day to restore my data and I'll eat my monitor with a side of keyboard."

Herb Brosowsky, Owner
Northwest Cheese

"If ever or ever a whiz there was,
The wizard of Tae, is one because,
Because, because, because, because, becuzzzzzzzz . .

I (you) saved my sale! Thank you Tae!"

Howard Petrick, Owner
Noe Valley Computers

"TSG: Even after all this time working together, I am still thrilled every time I interact with one of your people, either technical or admin. What I like best about you guys is that you LISTEN before you ACT so I know you are working on our REAL issues.

Last year's budget cuts could have hurt a lot worse but working with us to consolidate and clean up our database saved our organization big money. Thanks guys. We hit the jackpot with you!"

Neely Bogad, Executive Assistant/Board Liaison
The Contemporary Jewish Museum (Non-Profit)

"Thank you TSG for turning a hodge-podge of old equipment into a fully functional, fast and efficient system. With all of the different social programs we manage in so many physical locations, it was mind-boggling for us to handle this internally. Now our programs can grow and work toward our mission of reducing poverty, supporting families, and empowering communities.

Your concern for our welfare has always been genuine. I thank you, our staff thanks you and the one up there does too."

Joe Karingida, CFO
Catholic Charities

"You rock. You have saved my bacon so many times with extra hands and extra brain-power. As you know, this job is way bigger than one person (so says me, not the powers-that-be). Getting the 411 system finished is a real feather in my fedora and TSG should get some of the accolades. Don't try to hide--if I ever blow this pop stand I will find you. Sincerely guys, thanks for all you have done."

Sean Myers, Senior IT Engineer
Pacific Bell

What People Are Saying About TSG More Recently . . .

"TSG was an invaluable partner when my school district began to move into the digital age. We were a small district, with no internal technology expertise, just a few school level computer techs and self-trained data processing clerks. We had a mix of platforms and no systematic infrastructure. TSG made a long term commitment (over 15 years) working side-by-side with us to develop the technology to support our vision.

I appreciated the way Gloria, Teryl, Herb and their staff became part of our team—they were working for us, not trying to sell us something. They participated in meetings with end-users, many of whom resisted change, to better understand their perspectives . This personal contact was crucial in building our staff's trust in them , acceptance of change, and confidence to execute the new approaches.

Their honesty, careful planning, objectivity, and collaboration were essential to helping us move from no internal system to what is now a robust sustaining operation. Each year, as we worked through our multiyear plans, Gloria and Teryl were open to adjusting the types and intensities of their support as we built our internal capacity. They did not attempt to create dependency, as so many vendors would, but engaged in work that increased our district's staff capacity and independence. TSG brought access to a breadth of expertise and up-to-date information about technology changes that allowed us to anticipate future developments rather than being saddled with outdated systems.

It is also rare to find tech experts who are also very effective "people" persons –TSG has it all."

Christine Anderson, Ed.D.
Former Assistant Superintendent, Educational Services
Tamalpais Union High School District

"Herb and Teryl - Just a quick but heart-felt note of thanks to you both for all you, Tae and Carlos have done for PRC. Mountains have been moved and we are tremendously grateful! Talk with you soon. – Ann"

Ann Cooney
Pacifica Resource Center (Non-Profit)

*"Gloria! I am here in Yuba City....typing away on a thin client connected toVirmedice! Gloria, **your guys are top notch**. Everything is so tidy and clean (even for an OCD like me!). The 'anti-server' room/closet looks great. They just went to lunch, but I just wanted you to know that everything is going great and both helpdesk issues we had this morning were resolved by them in a matter of minutes. Kudos.*

Now for the fun part...connecting everything to the Konica printer, getting the scanners to work, getting Dr. Abouesh's laptop configured, etc. etc.."

Teresa Letlow, Director of Marketing
Therapeutic Solutions

"Gloria is a fantastic resource for business guidance and serves as a member on my Business and Technology Advisory Board. Her wealth of knowledge has provided me with a framework for several successful business endeavors." February 25, 2010

Matthew Zaroff, President
NetOps Corporation

"Gloria, Herb and Teryl; It's hard to imagine this "situation" is beginning to feel calmer now, in control and getting adjusted at last! It was hard to imagine that threshold of achievement would come after over two years of implementation, discovering and working through endless issues of program and network management and staff training, including my own! But you said it would! And how right you were! How does a high school district implement a new web-based, all-staff inclusive, student database program? Not without the expertise, and professional and personal integrity of our district's Network Management team at TSG Networks.

Thank You doesn't come close. We could not have done this without you. It's a pleasure working with you all and I'm sure we'll continue to have LOTS more fun together! ☺

My heartfelt best to you all, Connie

**Connie Ducey, Senior IT Data Specialist,
Tamalpais Union High School District, Larkspur, CA**

"Teryl, I love being in your network, since you are basically responsible for my entire career. Think about it, I might have ended up a line cook (no disrespect intended—you know how much I love to cook). Thank you for that. Seriously. How could I ever repay you? How could I have raised a family & bought a house without the jumpstart you gave me? Isn't life weird?"

Former employee

"Herb et al, Thank you all so much for the beautiful Birds of Paradise flowers. When they came on Friday, I thought, who would send me flowers, it's not my birthday. What a nice surprise! The best thing about this is that we are always there for each other. It is great when you can do business with people that you like and admire. It is the icing on the cake."

*Thanks again,
Rosalie and Cagney and Lacy*

**Rosalie Bulach, Retired President
Name-Finders Lists, Inc.**

"Hi Gloria & Teryl, . . . in our case as a "one man shop" I do not have peers to recognize good work or to complain about users with (smile). It's only when there is trouble or things break that I hear about it. I imagine that your experience is similar and I feel strongly that good work or service should be recognized.

I would especially like to commend Carlos and TSG for the way in which the service outage was handled while I was out on vacation. That was a major event which was kept to a minimum amount of downtime and service interruption thanks to everyone's efforts. I would like you to know that Carlos has been doing an excellent job here and that I am extremely pleased so far with his work and our relationship with TSG. Thank you,"

Scott Gordon, Information Technology Manager
San Francisco SPCA

"While working as the in-house Network Admin, I had the opportunity to work with Gloria and her staff at TSG for almost 10 years doing our Server and Network Maintenance. In that entire time, our network of almost 100 users (and 14 servers) had almost no down time during business hours and I'm proud to say our servers were virus/spyware free. All the Techs are extremely knowledgeable, always helpful and came up with creative ideas to link our remote offices and allow secure off-site access using Citrix. I would definitely hire them again and

recommend them to small businesses as well as larger ones

Roe Tyler,
Network Administrator
California Land Title of Marin

Routine maintenance? Ok, ok, ok! Jeez, some people. If you knew what was on my plate, you would understand.

Friday will work. And thanks for being a pain in my #@*! Regards, Todd

Todd Noah,
Intellectual Property Attorney
Dergosits & Noah, LLP

Chapter 16

Useful Templates

(Electronic versions:
www.tsgnetworks.com/templates)

Sample: Network Documentation

This is a possible **Table of Contents** you can use to
create a living documentation set for your company.
You can add or delete sections to fit your circumstances.

ABC Company
Network Documentation
Last Update: _____   By:_____

SEC 1:   Telephone Book - Keep all of your important
vendor and service provider phone numbers here.

SEC 2:   Network Diagrams - Update this section if you
change anything on your system.

SEC 3:   Network Configurations - Update this section if
you change anything on your system.

SEC 4:   Disaster Recovery Plan - Includes an analysis of
your current disaster recovery abilities, and
information on your backup equipment and power
protection devices.

SEC 5:   Cable Plant - Contains specific site locations and
station numbers. Include any updates to the
cabling plant and summaries of test analysis.

SEC 6:   Service Provider Information & Agreements –
Insert copies of all service and support agreements
from vendors.

SEC 7:   Software Licensing Log – Record existing software
license information and amend as software
licenses are renewed or new software is added.

SEC 9:   Site Visit Log – Contains summary information
regarding onsite visits by outside vendors for
historical reference.

SEC 10: Notes – Use this section to store information
temporarily while servicing your network.

Sample: Internet Usage Policy

ABC Company
Acceptable Internet Usage Policy

Use of the internet by staff members of **ABC Company** is permitted and encouraged where such use supports the goals and objectives of the business. All Staff must ensure that they:

Comply with current legislation; use the internet in an acceptable way; do not create unnecessary business risk to the company by their misuse of the internet.

Unacceptable Usage:
In particular the following is deemed unacceptable use by staff members:

- visiting internet sites that contain obscene, hateful, pornographic or otherwise illegal material
- using the computer to perpetrate any form of fraud, or software, film or music piracy
- using the internet to send offensive or harassing material to other users
- downloading commercial software or any copyrighted materials belonging to third parties
- hacking into unauthorized areas
- publishing defamatory and/or knowingly false material about **ABC Company** , your colleagues and/or our customers on social networking sites, 'blogs' (online journals), 'wikis' and any online publishing format
- revealing confidential information about **ABC Company** including financial information and information relating to our customers, business plans, policies, staff and/or internal discussions
- undertaking deliberate activities that waste staff effort or networked resources
- introducing any form of malicious software into the corporate network

Company-owned information held on third-party websites
If you produce, collect and/or process business-related information in the course of your work, the information remains the property of **ABC Company**. This includes such information stored on third-party websites such as webmail service providers and social networking sites, such as Facebook.

Monitoring
ABC Company accepts that the use of the internet is a valuable business tool. However, misuse of this facility can have a negative impact upon productivity and the reputation of the business.

In addition, all of the company's internet-related resources are provided for business purposes. Therefore, the company maintains the right to monitor the volume of internet and network traffic, together with the internet sites visited.

Sample: Internet Usage Policy, Continued

ABC Company
Acceptable Internet Usage Policy, Continued

The specific content of any transactions will not be monitored unless there is a suspicion of improper use.

Sanctions
Where it is believed that a staff member has failed to comply with this policy, they will face the company's disciplinary procedure. If the staff member is found to have breached the policy, they will face a disciplinary penalty ranging from a verbal warning to dismissal. The actual penalty applied will depend on factors such as the seriousness of the breach and the staff member's disciplinary record.

Agreement
All company staff members, contractors or temporary staff who have been granted the right to use the company's internet access are required to sign this agreement confirming their understanding and acceptance of this policy.

Staff Member acknowledgement:

I have read and understand these policies: _____

Date: _____

I witnessed the acknowledgement above: _____

Date: _____

Sample Template:

**Cost of Unplanned Down Time –
6 Person Business**

Average salary per employee	$50,000
Benefits at 20%	10,000
Total Ave Annual Salary/Employee:	**$60,000**
Work Hours in One Year	2,080
Total Average Hourly Salary/Employee:	$ 28.85
Number of Employees	6
Total Payroll Cost Per hour:	$173
Payroll Cost, wait hours tech to arrive (2)	$346
Payroll Cost, tech is working on issue (3)	$519
	$865
Hourly rate for technician @ $150/HR	$450
Travel time charge for technician	$ 65
Total Cost of technician for 3 Hours	**$515**

**Total Cost of Unplanned, 3 Hours of Server
Down Time: $ 1,380**

THE AUTHOR

TERYL L. BURT

Teryl is the middle-kid who along with her siblings Gloria and Herb, founded their first technology business in 1979. Within the first six years, their company earned a place on the **INC 500 Fastest-Growing Private Companies in America** list, showing 1,602.3% growth over five years.

Today, their company, **TSG Networks**, provides Cloud integration services and managed services, that are designed to transfer the burden and expense of routine management and maintenance of the Client's IT infrastructure to TSG's professional IT team, at a fraction of the cost.

Teryl lives in El Cerrito, CA. She has a BS in Education from SIU, a certificate in Project Management from University of California Hayward and is a 2006 graduate of the IPED Channel Elite MBA program.

About TSG

Leveraging almost a century of combined experience, TSG is committed to finding the right solution for each Client, based on individualized attention. We believe that everyone can streamline their systems, increase performance and still manage to contain costs. It takes an experienced team that has successfully executed before, complete dedication to make it happen, and the patience and determination to find the best fit.

TSG's diverse Client base is made up of small and medium sized businesses and medical practices headquartered in and around the San Francisco bay area, many with remote offices around the world, non-profit organizations, and schools of various sorts and sizes.

Determination, the founders know. It never quite came to fisticuffs growing up but thanks to a feisty mom ("go fight it out and come back when you have reasoned out the answer between yourselves"), they find the solution, for each and every Client, each and every time. When the needs of TSG's Clients change, TSG changes to meet them.

What TSG Does Better

We Respond

Chances are good that your call will be answered by a live person every time you call. Whoever that is will know how to further the process of getting you the result you want.

We Communicate

Because we focus on your business instead of technology, we talk your language and keep the Geek Speak in our lab (unless you are interested in the details). We work to address each Client's unique business situation and develop the communications method that best fits the business.

We Follow Through

If our Clients are satisfied, we grow. Growing your business by following through on each and every project, request or service initiative is our business.

WHAT DO *YOU* NEED TO KNOW TO MAKE YOUR BUSINESS MORE PRODUCTIVE?

If anything in this book hits a hot button, *it is time to consider your options.*

Start with a **No Geek Speak** business discussion about choices that can help your business, organization or school district achieve its goals.

When you are ready to figure it out, call us.

510.525.6210

www.tsgnetworks.com

Mention this book for a no-obligation consultation on the issue of your choice.

TSG NETWORKS

CLOUD INTEGRATOR

MANAGED SERVICES & TECHNOLOGY CONSULTING

www.ingramcontent.com/pod-product-compliance
Lightning Source LLC
Chambersburg PA
CBHW072159280526
45788CB00002B/803

* 9 7 8 1 4 5 6 3 7 8 6 0 8 *